The Mediterranean

The Mediterranean Sea, which laps the shores of such exotic places as the French Riviera, Egypt and the Greek Islands, holds some magnificent treasures for scientists and geologists, beachcombers and historians. This book will take you on a magical journey around the sea, and tell you about its importance to the people living in the very different countries surrounding it. It will explain how the Mediterranean was formed by the movement of continents and violent volcanic eruption over millions of years; show you some of its colourful and fascinating sea creatures; and discuss the dangers of pollution. You can explore ancient underwater cities, bustling modern ports, bleak volcanic islands and the vitally important Suez Canal. You will also learn about the natural resources of the sea and the scientific research now being carried out to unravel its mysteries. At the back you will find a glossary and a list of books to read.

SEAS AND OCEANS

The Mediterranean

Edited by Pat Hargreaves

WAYLAND

SILVER BURDETT

© 1980 Copyright Wayland Publishers Ltd
First published in 1980 by
Wayland Publishers Limited
49 Lansdowne Place, Hove
East Sussex BN3 1HF, England
ISBN 0 85340 767 3

Published in the United States by
Silver Burdett Company
Morristown, New Jersey
1980 printing
ISBN 0 382 06468 2

Phototypeset by
Trident Graphics Limited, Reigate, Surrey
Printed in Italy by
G. Canale & C.S.p.A., Turin

Seas and Oceans

Three-quarters of the earth's surface is covered by sea. Each book in this series takes you on a cruise of a mighty ocean, telling you of its history, discovery and exploration, the people who live on its shores, and the animals and plants found in and around it.

The Atlantic
The Caribbean and Gulf of Mexico
The Mediterranean
The Antarctic
The Arctic
The Indian Ocean
The Red Sea and Persian Gulf
The Pacific

Contents

1 A JOURNEY THROUGH THE MEDITERRANEAN

Above Holiday-makers enjoy sun, sand and sea at a popular resort in the south of France.

Opposite Village square, Malta. Farmers still use donkeys and carts in many parts of the Mediterranean.

Two thousand years ago, the Romans called the Mediterranean Sea *Mare Nostrum*, which is Latin for 'Our Sea'. At that time, all the countries bordering the Mediterranean were part of the Roman Empire.

If we could look down on the Mediterranean from an aeroplane we would be able to see a long narrow basin with the Straits of Gibraltar to the west. At its eastern end, the Mediterranean is enclosed by countries and by the shores of the Black Sea. Generally, the northern coasts of the Mediterranean have rainy, mild winters and hot summers. Fruit such as oranges, grapes, olives and figs are grown. The southern coasts are close to the Sahara Desert and are hot and dry. The countries of the south used to be poor but now that oil and gas have been discovered in Algeria and Libya, they are richer.

Scientists now know a great deal about the sea floor, the plants and animals that live in the sea and on the shore, the ancient undersea towns, and the movement of the water. In this book you will find out about all these interesting things, but first let us take a tour of the Mediterranean Sea, which is really several deep basins linked by narrow straits. We will visit each basin in turn from west to east. You can see them on the map.

The Straits of Gibraltar are only about sixteen kilometres (10 miles) wide. All the water of the Mediterranean comes in from the Atlantic Ocean through these Straits, or from rivers, or falls as rain. There is no connection

Above View across the Aegean Sea from the islands of Santorini.

Above The smoking crater of Stromboli. This volcano is one of the most active in the world.

with the other great oceans of the world except by the narrow Suez Canal which lies to the east.

To the east of Gibraltar is the Alboran Sea with the tiny island of Alboran in the middle. Then we come to the Balearic Basin which is bordered by France, Spain and Algeria. The islands of Sardinia and Corsica lie to the east. The sea to the north of Corsica is sometimes called the Ligurian Sea. To the south, between Corsica, Sardinia, Italy and Sicily we find the Tyrrhenian Sea, which in the middle is over 3,000 metres (9,840 ft) deep. The Tyrrhenian Sea is surrounded by several active volcanoes. Some of these are on little islands. The island of Stromboli has one of the most active volcanoes in the world. It erupts every twenty minutes or so like a huge firework.

The Straits of Messina and the Sicily Channel, which are both narrow and shallow, join the water of the western Mediterranean with that of the eastern end. To the north is the Adriatic Sea, which is long, narrow and mostly shallower than 200 metres (656 ft). During the Ice Age, when the level of the seas dropped, it was completely dry. The Italian coast of the Adriatic has sandy beaches, the Yugoslavian side is very mountainous.

The largest part of the Mediterranean Sea stretches from the island of Malta in the west to Lebanon, Syria and Israel in the east. The part between Italy and Greece is called the Ionian Sea, and that between the island of Crete and Libya is called the Libyan Sea. The area bordered by Turkey, Egypt and the

Middle Eastern countries is called the Levant Basin. The south-east corner of this basin is covered with sand and mud brought down by the River Nile.

There are many beautiful islands near Greece and Turkey. Some, such as Crete, have trees, grass and flowers. Others have little rain and are dry and rocky.

All around the Mediterranean you will find ancient cities. The oldest ports and harbours in the world are on its shores, and many ancient wrecks lie on the sea floor.

Below Fishermen drying their nets – a scene as ancient as man's presence on these shores.

ATLANTIC OCEAN

Ligurian Sea

Tyrrhenian
Sea

Balearic Basin

Straits of Gibraltar

Alboran Sea

Str

Sicily Cha

BLACK SEA

Aegean Sea

Sea

Ionian Sea

Straits of Messina

Libyan Sea

Levant Basin

Suez Canal

Red Sea

2 HOW THE SEA WAS FORMED

Continental drift

The continents are known to be lighter parts of the Earth's crust which float on a much denser, or heavier, lower layer of rock covering the whole globe. Continental landmasses have moved about over the surface of the earth for millions of years. Geologists call this 'continental drift'. Where continents have broken up and drifted apart, new oceans are formed between them.

The Mediterranean Sea is all that remains of a huge ocean that existed 150 million years ago. This ocean stretched from east to west between two former continental landmasses. Europe and Asia formed the landmass to the north. To the south lay Gondwanaland. This super-continent split apart and became our present-day Africa, India, Australia, Antarctica and South America. The story of how the Mediterranean formed is the story of the gradual disappearance of the great ocean known as Tethys, which lay between Gondwanaland and Eurasia.

When the Gondwanaland landmass to the south of the Tethys Ocean split up and formed the southern continents as we know them, they left in their wake the Indian Ocean. At the same time, this drifting caused the destruction of much of the eastern Tethys Ocean. India and Eurasia collided. The rocks of the sea floor between these two continents were thrust upwards and formed the great Himalayan mountain chain. Further west, movements of

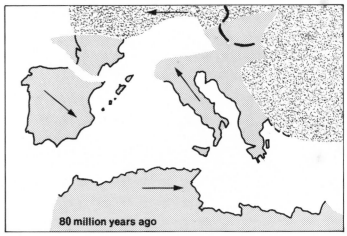

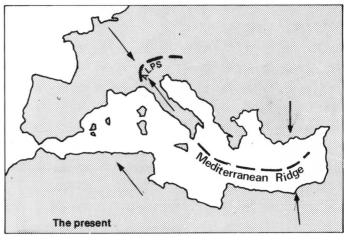

Above About 80 million years ago the movements of Africa and Italy towards Eurasia crumpled the ocean floor of the Tethys Ocean in the north-west, and formed the Western Alps. This movement continues to the present day. The ocean floor is being folded under the eastern Mediterranean along the Mediterranean Ridge. This explains why there is still so much volcanic activity in this area.

Far right The Alps are some of the wreckage that resulted when Italy collided with Europe.

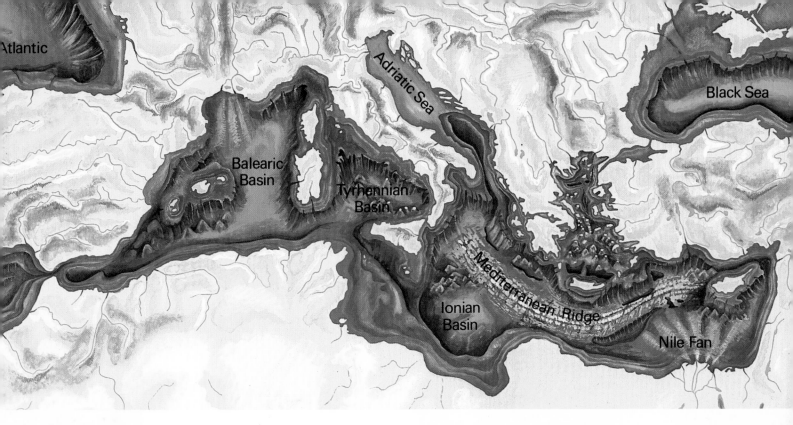

Balearic
Basin

Tyrhennian
Basin

Ionian
Basin

Mediterranean Ridge

Nile Fan

Atlantic

Adriatic Sea

Black Sea

Above If the Mediterranean was drained of its water, you would be able to see these mountain ranges, ridges and deep basins which make up the sea floor.

Africa and Italy towards Eurasia crumpled the ocean floor and formed the Alps.

The eastern Mediterranean Sea is all that remains of the mighty Tethys Ocean, but even now below its waters the slow destruction of the ocean floor goes on. A mountain chain known as the Mediterranean Ridge, is now being formed under the sea in this area.

The present-day deeps of the western Mediterranean basins are pieces of a new ocean floor that developed after the Western Alps were formed. The basins were created 25 million years ago when small blocks of continent (now forming Corsica, Sardinia and Sicily) moved away from the European landmass.

13

Earthquakes and volcanoes

Above This old man is sitting amongst the ruins of his house, after an earthquake shook his village in the hills outside Rome.

If you were to draw a map of the places in the world where earthquakes and volcanic eruptions have been recorded, the Mediterranean would quickly show up as a very active region. This is not only because most of the ancient civilized people lived here and recorded what happened, but also because this area has a very long history of earth movements.

When continents collide, they frequently thrust over or under one another. Major thrust zones beneath the Alpine and Himalayan mountain chains are the sources of many earthquakes – those of the Middle East, for example, which we read about in the newspapers. Earthquakes are also common where one piece of ocean floor is forced under another. This is happening below the island of Crete and under the 'toe' of Italy.

Zones of weakness and movement in the Earth's crust provide routes for hot rocks deep in the earth to find their way to the surface. Chains of volcanoes are strung out in arcs under the sea west of Italy and north of Crete. The first of these arcs contains some of the world's best-known volcanoes: Mount Etna on Sicily, and the island volcanoes of Stromboli, Lipari and Vulcano which sit on the sea floor with only their tips above the sea surface. Many of the islands of the Aegean Sea are also volcanoes or fragments of old explosive volcanoes.

Many zones of weakness in the earth's crust remain after the continental drift movements. Mount Vesuvius, near Naples, is on such a

Above Mount Etna is another volcano which is very much alive. A serious eruption occurred in 1979.

zone. Records of its eruptions go back to the earliest Mediterranean civilizations, but its most famous eruption occurred in Roman times in A.D. 79, when clouds of glowing volcanic ash engulfed the cities of Pompeii, Herculaneum and Stabiae. The ash cooled, and preserved in rock and dust are streets, beautiful buildings and even people just as they were when the ash buried them.

Right Artist's impression of the last hours of Pompeii before the town was buried by volcanic ash.

15

Rocks and sediments

The sea floors of Mediterranean basins are covered with sediments that in some respects are similar to those found in the deeper oceans. Almost all are deep-water sediments lying several thousand metres below sea level, since shallow 'continental shelf' regions along the coastline are rare in the Mediterranean. The sediments are partly made up of soft 'oozes', the skeletons of tiny animals and plants which, when alive, inhabit the surface waters. But, because land is close, Mediterranean sea-floor sediments are mixed with fine material such as silt and clay from the rivers, or dust blown in by winds from the deserts or from volcanoes.

Close to volcanic eruptions, layers of

Above Ruins of the Minoan palace at Knossos on Crete, possibly destroyed by a gigantic earthquake.

coarser volcanic ash are spread over the sea floor. As the pile of sediments mounts up with time, these layers are left to record past volcanic eruptions. A group of islands in the Aegean Sea, known as Santorini, is all that remains of a huge volcano that disappeared in a gigantic explosion about 3,500 years ago. An ash layer, resulting from this explosion, is frequently found when scientists sample the upper ten metres (33 ft) of sediment in the eastern Mediterranean. The discovery of this thickly spread layer of ash explained to archaeologists why towns of the Minoan civilization, which flourished in the Aegean in about 1500 B.C., suddenly died out.

There was one short period when Mediterranean sea-floor sediments were not of the deep-water type. Working in the deepest parts of the Mediterranean, the research drilling ship *Glomar Challenger* recovered samples of a rock formation that must have been laid down in a desert-type country with salt lakes. Rocks either side of this formation are of deep-water type. It is now thought that this peculiar sandwich of rock formations is due to the fact that there was a period of one million years when the Mediterranean repeatedly dried up (evaporated) and then flooded again, perhaps as many as twelve times. This was caused by the narrow strait which brought water from the open ocean being opened and closed by movements of the earth.

Left The islands of Santorini are largely made up of black volcanic rock.

Ancient peoples

The first cities in the world were built in the countries around the eastern Mediterranean just over 10,000 years ago. There was more rain then, and more forests and grass. In Turkey and Jordan, people who had lived in caves and hunted for fish and deer and eaten the seeds of wild grasses, first began to plant grass seeds in order to grow grain. They also started to keep animals such as dogs and goats, and to breed them so as to get the best use out of them. This was the beginning of farming, or agriculture.

Once people had settled on the land and started farming, it was natural that they should build solid houses out of mud bricks and wooden beams. Then they built many houses close together so that some people could specialize in making pottery, some could make cloth, and others could make axes and farm implements. So people began to exchange goods, although they had not yet invented money. The first towns were very small, only a few hundred or a few thousand people. Nowadays, even a small town has 50,000 people in it.

Eight thousand years ago, the first ships were sailing across the Aegean to collect a very useful kind of glassy stone called obsidian. Four thousand years ago, sailing ships were criss-crossing the eastern Mediterranean between the mainland and the islands of the Aegean Sea, or sailing along the coast to Egypt and Turkey. Two thousand years ago, the

Above The Island of Delos lies at the centre of a group of islands called the Cyclades in the Aegean Sea. In Greek times it was a sacred island dedicated to the god Apollo. No births or deaths were allowed on it, so a neighbouring island was used when these seemed likely.

Above A Roman war galley leaves port to patrol the empire.

Romans united all the countries around the Mediterranean in one empire, and at that time there were over 1,000 ports supplying the ships that sailed the Mediterranean.

Ancient ships were quite small, and their sails were not very efficient. Storms often drove the ships off their course, or on to rocks. The wooden ships sometimes sank under the weight of their cargo, and divers today find the remains half buried by sand and mud.

Later history

In the fifth century A.D. the Roman Empire collapsed, though much of its power shifted to its eastern branch of Byzantium (now Istanbul in Turkey) which was to survive for another thousand years. However, a powerful new force was unleashed in the East with the rise of Islam in the seventh century. Arab armies swept across the lands of the eastern Mediterranean and North Africa and crossed the sea into Spain. Their advance was finally checked in central France by Charles Martel and his Frankish knights in 732. It was many years, however, before they were driven from their hold on Spain, and the effects of their long

Above The naval battle of Lepanto, in which the Turkish fleet was defeated in 1571.

Above The walled city of Constantinople, capital of the Byzantine Empire.

Left This medieval painting shows a Christian fleet attacking a Turkish stronghold in southern Italy.

occupation can still be seen on the peoples and culture of that country today. Many Arabs took to the sea as Barbary pirates, much feared right up to the nineteenth century.

Alarmed by the conquest of the Holy Land, the Christian rulers of Western Europe began a series of crusades, and for a short time established a kingdom at Jerusalem. Though the crusades did not succeed in the end, they did open up trading and cultural links between the East and West. The remnants of the Byzantine Empire fell to the Turks in 1453, but

their expansion in the Mediterranean was halted by their defeat at the naval battle of Lepanto in 1571.

Britain's naval interest in the Mediterranean began with her brief possession of Tangier in the seventeenth century. It continued through a series of wars with France ending in 1815, and then more peacefully until the two world wars of this century. Two of Britain's main concerns were possession of the valuable trade with the East, and protection of the route to India, particularly after the opening of the Suez Canal in 1869. The capture of Gibraltar in 1704 provided a British naval base, supplemented later by Menorca and then Malta.

During the Middle Ages, map-makers produced decorative charts, known as portulans, for Mediterranean seafarers. Modern charting began in the nineteenth century.

4 TOWNS BENEATH THE SEA

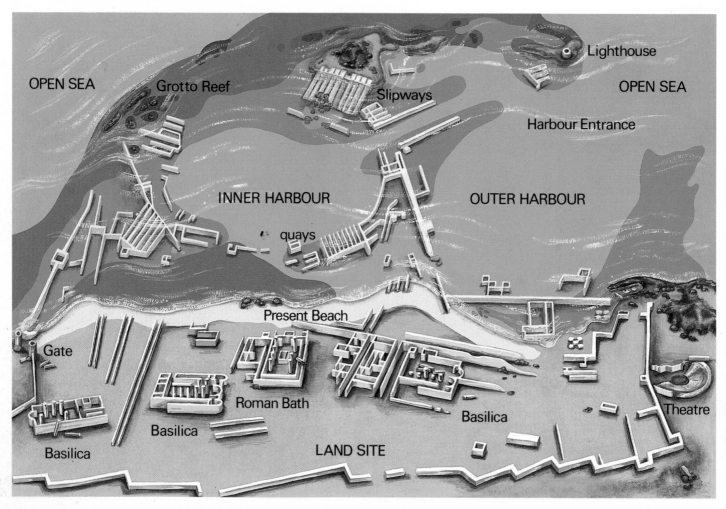

Above The ancient Greek harbour of Apollonia, showing ruins on land, and buildings now under the sea.

If you told people that you had seen buildings and streets under the sea, they would probably laugh at you. But truth is sometimes stranger than fiction. Around the shores of the Mediterranean, volcanic eruptions and earthquakes occur every year. When there is an earthquake the land may rise or sink, sometimes by as much as a few metres, and buildings are shaken or knocked over. If there is an earth-

quake near the sea the coast may sink or come up out of the sea.

Recently, archaeologists, divers and geologists have studied ancient harbours on the coast of the Mediterranean. In the places where there have been no earthquakes, they find that the old Roman and Greek ports of 2,000 years ago are on the shore just where one would expect. Where there *have* been

THE MOVING WATER

eat, salt and weather

erybody's picture of the Mediterranean Sea f clear skies, blue sea and hot sunshine. A nate ideal for holidays. However, the ther is not always so good, and the seasons e a great effect on the waters of the diterranean.

The only connection to the Atlantic Ocean is ough the narrow Straits of Gibraltar be-en the southern tip of Spain and the coast of rth Africa. The Atlantic water flows in at the face, and inside the almost land-locked diterranean its temperature and salinity tiness) are altered. More fresh water is porated from the surface of the sea by the and wind than enters it from the rivers, and he upper layers of water become warm and y. (Put a saucer of salt water outside on a day and see how it gets saltier as the water porates.)

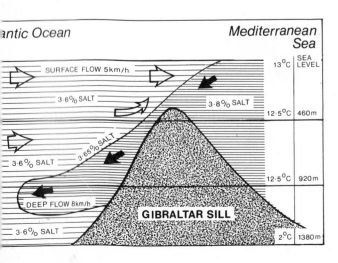

In winter, the weather in the Gulf of Lions in the South of France can become surprisingly cold and stormy as the Mistral wind sweeps down from the Pyrenees. As well as making this sea area very rough it has an effect on the previously warm, salty water. The winds cool the sea surface and evaporate even more fresh water. This makes the surface layers denser (heavier). Cold water is heavier than warm water and salty water is heavier than fresh water. These changes happen to such an extent that the surface water sinks to the sea bed 2,000 metres (1 mile) below where it joins water which was cooled in previous winters. This cool, salty water fills the bottom of the western part of the Mediterranean and flows out into the Atlantic at Gibraltar below the incoming Atlantic water.

The water from the Mediterranean is different both in temperature and saltiness from that of the Atlantic Ocean. It spreads out at a depth of around 1,000 metres (half a mile) and can be found in the Atlantic as far north as Scotland and as far west as the island of Bermuda.

Left A cross-section of the sea which shows the flow of water in and out of the Mediterranean at Gibraltar. Notice that the temperature is higher at the surface. The Mediterranean is slightly saltier than the Atlantic.

Right Salt pans in the south of France. Sea-water is channelled into pools where the heat of the sun evaporates the water, leaving salt behind.

Overleaf Storm clouds gather over the western coast of Crete.

earthquakes they sometimes find cities under the sea, or harbours lifted out of the water and quite dry. First we shall see where the uplifted harbours are, and then we will look at some of the sunken towns.

Most of the harbours which are now out of the sea are on islands between Greece and Turkey. Here are some of their names: Potamos, on the island of Antikythera; Phala-sarna, Sulia and Musagores on Crete; and Calithea, Afandea and Lindos on Rhodes. The biggest and best-preserved underwater cities are Pavlo Petri in southern Greece, Apollonia in Libya, and Misenum, Baiae and Possilipo in

Above This Greek statue of two young girls was found off the coast of Italy.

Italy. There are many hundreds more. We shall look at Pavlo Petri, Apollonia and Misenum.

Imagine that you are swimming with a mask and flippers in the blue water. You are looking at the sea floor. There are stones, rocks and sandy patches. Suddenly you see a large rock with straight-cut edges. You swim over it and see a corridor leading to a square room. The roof has fallen in, slabs of stone and broken pottery lie on the floor. Further over there are other oblong shapes like houses with rooms. In some of the rooms there are stone boxes about one metre (3 ft) long. There are lines of stones such as you might find on the edge of a road. This is how the Bronze Age town of Pavlo Petri looked when it was discovered in 1967. Later, archaeologists and divers explored it as well. They found that it was built 3,500 years ago. They measured the buildings and roads, and they found a bronze statue and beautiful pottery. They discovered that the stone boxes were really graves with skeletons in them.

Apollonia was founded 2,600 years ago by

people from Greece who settled in Libya the city is 2.5 metres (8 ft) under Originally three islands enclosed a h with slipways, so that ships could be pull of the water for repairs. There were doc quays around the harbour, and a light About 1,400 years ago the city sank slo the sea. Now divers can swim easily o stone slipways and quays.

In the Bay of Naples there are nine water cities. If you wear diving equipm look carefully, you can find the old Rom from Naples to Misenum which h underwater because of earthquakes. enum, divers have found square tan the rocks which look like swimming p Romans used to keep live fish in th they wanted to eat them. There refrigerators in those days.

These are just some of the stran that have been found in underwater t careful exploration of these ancient are beginning to understand how and were used.

Left We can learn much about daily life in Greek times from the scenes which artists used to decorate their pottery. On this vase boys are playing ball.

Measurements in the water

Scientists have invented ways to measure the temperature and saltiness of the oceans. The first method developed was to lower a series of 'water bottles' on a wire into the sea below the ship. These bottles were fastened to the wire at the level at which scientists wanted to take measurements. A weight called a 'messenger' was then sent sliding down the wire, and as it passed each level the bottle trapped some sea-water and, at the same time, sensitive thermometers measured the water temperature. The bottles were then pulled back to the ship and the thermometer reading recorded. The saltiness of the water used to be measured chemically, but now it is found by seeing how well the water conducts electricity. The saltier the water the better conductor it is.

New electronic instruments can measure temperature and saltiness automatically and send detailed information back to the ship through an electric cable. Satellites can also be used to see the changes in temperature of the ocean surface over large areas.

Scientists can now measure currents in the deepest parts of the ocean. Some measurements are made with recording meters. These use propellers to measure how fast the water flows, and a vane and magnetic compass to find direction. The information is stored on a small tape recorder inside the instrument.

Above An echo-sounder is towed behind a research ship to chart the sea floor.

Left 'Water bottles' like this one are used to measure the sea's temperature and saltiness.

Another method is to use a float which moves with the water at a fixed depth. Each float is able to transmit sound signals which can be picked up by the research ship. In this way scientists are able to track the floats as they are taken along by the currents at all depths. The lightly weighted ones float near the surface and the heavier ones float nearer to the sea floor.

Right Floating buoys are used to record wind speeds, temperature and currents.

Waves and tides

Mediterranean tides are small compared with those of most oceans and seas. Ocean tides are restricted by the narrow Straits of Gibraltar and the sea is too small for the gravity of the Moon and Sun to have much effect. Ignorance of tides caused difficulties for Julius Caesar and his ships when he first tried to invade Britain in 54 B.C.

'Tidal range' means the difference between the level of the water at low and high tides. Venice and Trieste in the northern Adriatic have tidal ranges of up to one metre (3 ft), but elsewhere ranges are only a few centimetres. Even so, tidal currents may be very strong. Through the Straits of Messina between Sicily and the Italian mainland, the speed of the currents sometimes reaches two metres (6 ft) a second. It has been said that the frustrated Greek philosopher, Aristotle, drowned himself in the strong currents of the

Euripus, off the Island of Euboea, because he could not explain their cause.

Although large tides elsewhere cause difficulties for shipping, the small tides of the Mediterranean cause difficulties too. Large vessels are unable to navigate far up the rivers because there are no tidal currents to keep the entrance free from silt.

Waves are usually small too, because of the limited area over which the winds blow. Sometimes the strong local winds of the Mistral, the Sirocco or the Bora produce large waves. Between Sicily and Tunisia waves can reach heights of more than twelve metres (40 ft). In the northern Adriatic the coast is exposed to large waves from the south-east.

For the low-lying, historic city of Venice, the greatest danger comes from the sea. The average sea level throughout the world is gradually rising by about ten centimetres (4 in) a century, due to the melting of polar ice-caps, and at the same time Venice is slowly sinking. When high tides are increased even further by the weather, especially by the Sirocco winds, the resulting *aqua alta* (high water) causes serious flooding. In 1966 many ancient monuments and works of art were badly damaged.

Left St Marks Square in Venice was under water for several days following high tides and bad weather in 1966.

Right A waterspout is a moving column of water caused by a whirlwind passing over the sea. This one occurred off the coast of Sicily during a heavy storm.

6 RESOURCES

Fishing

Each year, about one million tonnes of fish are taken from the Mediterranean by the seventeen countries surrounding these waters. This is not a very important amount of fish compared with the world catch of about 75 million tonnes, or with the Atlantic catch of 25 million tonnes. However, fishing is very important to the people of the Mediterranean, and it gives work to more than 300,000 fishermen. Some countries use very modern fishing techniques and are very efficient, while others use old-fashioned methods.

Most of the fish caught in the Mediterranean are herrings, sardines or similar fish which do not swim very deep. Sometimes fishermen attract them at night with powerful lamps fitted to their boats. When a big enough shoal of fish has been attracted, other boats surround them with a long net called a *lampara*. Another important fish is a red fish called a *rouget*, while less important are mackerel, cod, flatfish, tuna fish and shark. Sometimes tuna are caught in big nets made like a maze which are placed quite near to the shore. The tuna find their way in but cannot get out.

Shellfish are an important part of the diet of Mediterranean people. Besides lobsters, crabs and shrimps, there are mussels, oysters and clams. Squid and octopus are also a great delicacy in these countries.

Modern methods of fishing unfortunately enable people to catch fish faster than the fish can grow. It is only since 1970 that all seventeen nations have got together to try to save fish stocks from destruction. Two of the problems are that some fish-breeding areas have been filled in to make more land, and that pollution (see p. 42) is getting worse. Let us hope that governments will find methods to save the fish stocks before it is too late.

Below Shellfish like this clam are a great delicacy in Mediterranean countries.

Opposite Small boats like this one are used all over the Mediterranean to fish for shellfish, squid and octopus.

Other resources

The natural energy resources with which we have become more familiar are oil and natural gas. Since drilling began off the coast of Italy in 1960, more than fifteen oil or gas fields have been found and are providing fuel for Mediterranean countries.

Another resource is the sponge. The warm water of the Mediterranean provides the right conditions for sponges to grow. They live on the sea floor and feed by taking in water and nutrients through tiny holes in their body. Sponges are collected by divers, or by fishermen dragging rakes from boats. They are dried and the skeletons used as bath sponges.

Warm water also helps the growth of seaweeds like kelp. Some varieties grow very long indeed in the sea, and the top few metres are cropped by special boats. Other types of kelp are found along the shore. Seaweed is used in the manufacture of drugs, and for food additives, one of which is used in the manufacture of ice-cream and cream toppings.

On the hot, dry coasts of Israel and North Africa, sea-water itself is a resource. Special desalination plants are used to take the salt out of sea-water. The salt-free water is used for irrigation and domestic purposes. The salt in the sea is another resource. Sometimes it is dug from natural deep deposits dating from millions of years ago. Another method is to dry

Above Seaweed is gathered by the cartload on a beach in Mallorca.

out salt pans (shallow ponds of salty water) until a thick layer of salt is left.

It has been suggested that a great barrage or dam be placed across the Straits of Gibraltar to 'harness the tides'. The difference in water level at each side of the barrage could be used to turn great turbines which would produce electricity. Perhaps this will happen sometime in the future.

Left The best sponges in the world come from the eastern Mediterranean. This Greek boat is drying its haul in the sun before selling them on shore.

Overleaf Saltpans on Malta's rocky coastline.

Leisure in the Mediterranean

We have seen that the seas provide us with many resources, but probably the first thing you think of when the sea is mentioned is having fun. Most people enjoy a seaside holiday, and that does not necessarily mean just lying on a beach sunbathing. There are all kinds of interesting things to do in, on and around the sea and its shores.

For leisure activities, such as sailing, swimming and scuba-diving, the Mediterranean is ideal. For much of the year the sea-water is warm because the sun is hot and shines a good deal of the time. There are also plenty of beautiful beaches to explore for strange seaweeds, animals and shells. Many islands, such as the Balearics, Sicily, Crete and Sardinia, attract tourists, as do some of the larger countries which border the Mediterranean. Indeed, money spent by tourists is often very

important to a country's economy. Some countries need this money to help pay for goods that they have to import from abroad.

We have seen that the area is rich in history and that there are remains of some cities under the sea. Tourists with diving equipment can visit these sites. If the water is shallow and clear they may be lucky enough to see ancient remains of houses and harbours.

People are often fascinated by the beautiful fish which dart about in the sparkling water. It is possible to hire glass-bottomed boats from which you can see these fish easily.

Sailing is also very popular. Yachts have large sails which catch the wind and speed them over the waves. Some people have motor boats from which they water-ski. Some motor boats are large enough for holiday-makers to live on. There are also the huge cruise-liners specially built to take thousands of passengers around the Mediterranean. They often call at ports, so that people can explore the different cities. They can see how the people live, visit interesting museums, taste the local food and buy souvenirs in the markets.

If we want to go on enjoying the pleasures of the Mediterranean, however, we must stop polluting its waters.

Left Swimming, sunbathing, sailing and water-skiing are just some of the ways people can enjoy the sea.

Right Diver with an old anchor from one of the Mediterranean's many wrecks.

Pollution

Sea-water contains a large number of chemicals from natural sources. These include sodium, chloride, calcium and magnesium. There are also minute traces of many other elements. Some, such as silicon, phosphorus and oxygen, are essential to marine life. It is important to maintain the sea in its natural state and to avoid polluting it with dangerous chemicals, oil, sewage and other wastes which could endanger life in the oceans.

In the open ocean, pollution is not yet a serious problem because there are large amounts of water in which harmful chemicals can be diluted. The Mediterranean Sea is particularly vulnerable because the water is almost enclosed by land, and except where there are strong winds there are no large waves or strong currents to wash away pollutants to the open ocean.

Pollution can arise from a number of different sources. In this heavily populated area, sewage is a serious problem. Many countries still pipe it directly to rivers or to the sea. Sometimes, decomposing sewage and other organic matter cause the oxygen in the water to be used up so that marine life cannot survive. Industrial waste too is sometimes discharged into rivers or the sea. These chemical pollutants include lead, ferric oxide and acids – also some fertilizers used in agriculture. These fertilizers tend to drain into rivers, or are carried by the wind to the sea.

In the Mediterranean, however, the main danger comes from oil spillage. Sometimes

Above The tide brings in a sticky mass of oil to pollute a Mediterranean beach.

this occurs accidentally, but quite often oil tankers flush out their tanks with sea-water and the oil left in the tanks pollutes the sea. Recently, many of the Mediterranean countries have become more aware of the damaging effects of pollution and are taking measures to avoid it. In some areas scientists study the sea carefully to ensure that the amount of pollution does not increase and become dangerous to plant and animal life.

Left Oil pollution is often fatal to sea-birds.

8 THE SHORES OF THE MEDITERRANEAN

The shores of the Mediterranean cover a very large area and, as we might expect, the type of ground varies from hard rock to shingle, mud and sand. Rocky shores provide a good base for seaweeds (which are really collections of larger algae) and for animals which like to attach themselves to the rocks. Shingle beaches seldom have many plants or animals living on them because the waves grind the pebbles together, so that anything alive may be crushed. Sand or mud may provide homes not only for the animals which crawl over it but also for those which burrow down to shelter from predators. (Predators are animals which prey on other animals for food.)

On some ocean beaches, especially on the upper shore, the movement of the water due to waves or to tides has a great effect on life

Below Rocky shores provide a home for a wealth of sea-shore life.

Above Acorn barnacles are usually found on the lower shore cemented to rocks, wharves, boats or sometimes to the shells of other animals.

there. The Mediterranean is distinct from many seas and oceans of the world in having only small tides. Even so, plants and animals may be alternately covered by water and then exposed to air. This, together with heat from the sun in summer and other climatic conditions, makes it a harsh environment for shore life to tolerate.

Further down the shore, nearer to the sea, plants and animals may be covered with water more of the time. Different species (types) are adapted to tolerate different shore conditions, so that a glance at an ocean shore reveals clearly defined zones or areas.

What types of algae (seaweeds) are we likely to find? In the upper zones of the Mediterranean rocky shores there are green and red algae of many different kinds. Some are 'lacey'

45

Above Black-headed gull on nesting site. Gulls are common on beaches and around harbours where they quarrel amongst themselves for scraps of food.

and others have branches, but all are attached to the rocks by a special 'holdfast'. Among the plants, a few marine animals may be moving – perhaps crabs or even sandhoppers, which are related to woodlice. The algae help to provide shelter, and animals such as the periwinkle feed on them.

The lower rocky zone nearer the sea may have red, green or possibly brown algae, depending on the region. Other animals in this zone include limpets, barnacles or mussels attached to rocks. There are also hermit crabs and coat-of-mail shells. Hermit crabs have soft bodies, so they live in the shells of dead sea animals which they carry about with them for protection.

On sandy shores we may find sea grasses with a rooting system, but few algae. This is because the algae have no roots to anchor themselves in soft places. Occasionally we find algae attached to sea grasses, sea mats and sea anemones. In the sand there are burrowers such as lugworms and ragworms. Bivalve seashells, which look like cockles, may also be found.

So far we have said little about how all these animals feed. Attached animals such as barnacles and mussels filter tiny particles of food from sea-water as it washes over them. These and some of the plant-eaters (herbivores) such as periwinkles may be preyed upon by flesh-eaters (carnivores) – whelks and seabirds, for example. The most familiar birds of the shore are gulls. The black-headed gull is common in the Mediterranean. Together with other species of gull it scavenges for food, frequently accompanying fishing boats and feeding on the offal thrown over the side by the fishermen. Shearwaters, diving birds, ducks and puffins are also found in this area.

Right These snakelock anemones look like plants but are in fact animals. Their wavy, leaf-like tentacles are used to capture prey in the rock pools where they live.

Above Octopuses catch their prey with their long tentacles. This one is about to pounce on a crab.

floor. Imagine a journey down into the depths. Gradually as we go deeper the light decreases. From about 200 to about 800 metres (656 to 2,624 ft) we find ourselves in a 'twilight' zone. As we go even deeper there is inky blackness. In these layers there are more tiny planktonic animals such as copepods, shrimps, prawns, comb jellies, jellyfish, arrow worms and many others. We also might find large deep-sea fish, and squid with long tentacles. Many of the animals are darker in colour than those in the surface layers. The fish are often black or grey and the prawns dark red. We might see numerous tiny lights moving about. These are on the bodies of many of the animals including fish and squid. They make beautiful patterns and are often coloured a glowing blue.

What do the animals here eat? Herbivores move up into the surface layers and eat the plants, and some carnivores may also ascend and eat at the surface. It is well known that some of the plankton in the twilight zone move upwards at dusk and descend to the depths the following morning. Many of the carnivores, however, remain in the depths throughout their life.

Life on the sea floor

The sea floor near the coastline and islands of the Mediterranean is sometimes so shallow that sunlight can penetrate to the bottom and plants can grow. Algae, including long kelp, may be encrusted on rocks or there may be long kelp and sea grasses underwater in sandy areas. Far off shore, the water is very deep – 3,0000 metres (9,250 ft) in some places. Sunlight cannot reach down this far and it is very cold and dark. The water pressure there is very great.

Let us imagine that we could walk underwater along the sea floor a little way off shore where the sea is up to 300 metres (984 ft) deep. What types of animals would we find? We might find beautiful orange sponges, corals, sea anemones of many different colours, barnacles, sea mats, or purple sea squirts. A special type of pink coral there is collected for making jewellery. There may also be spiny sea urchins. We would find starfish creeping over the sand and mud and burrowing into it to feed on worms and shells. Some worms eat sand or mud to extract organic materials; others have beautiful tentacles with which they trap particles of food in the water. We would also find larger animals such as crabs, lobsters, fish or even an octopus. There might be flatfish, such as plaice, which lie quietly near the sea bottom, or the more dangerous carnivores like the vicious Moray eels which swim around the deep rocks, caves or shipwrecks.

Far off shore, at the bottom of the very deepest sea areas, there may be some of these

Above One of the many beautiful corals found in the Mediterranean.

Left Sea squirts are small bag-like creatures, usually fixed to the bottom. They have two openings in their bodies – one for taking in water and food and the other to 'squirt' out waste water.

animals, and others too. Their presence depends on many different factors, such as type of sediment on the sea floor and what there is to eat. Bottom-living animals are adapted to the great pressure and inky blackness. For instance, some fish and prawns there do not rely on their eyesight but use fins or feelers to help them find their way around.

Sicily

Many islands are scattered throughout the Mediterranean. They range in size from the largest (such as Sicily and Cyprus) to the smallest isles of Greece. A visit to the islands is really a visit to many different countries which have a variety of languages and customs. Yet these islands are similar too. Many are rugged, quite often with mountains, leaving only small areas of flat land suitable for farming and building.

Sicily is typical of one of the larger islands of the Mediterranean. Situated between the mainland of Italy and the North African coast, Sicily lies at the crossroads of the ancient trade routes. In the days of the Roman Empire it was called the 'Granary of the Empire', as grain was grown there. Later, Arab invaders brought prosperity to the island. They introduced irrigation and manufactured many different products from silk, leather, gold and silver.

Today, Sicily has a high population. There is very little modern industry, but some of the old craft industries, such as the making of leather-goods, continues. Fishing is important, though most is carried out by villagers in small boats. There is also some boat-building. Away from the coasts the landscape is dry and mountain-

Below The fertile slopes of Mount Etna provide some of the richest farmland in Sicily.

Above The grape harvest is one of the busiest times of year for the Mediterranean farmer.

ous. The rocks, which are mainly sandstone and limestone, do not form very fertile soils. Mount Etna is at the centre of an area of volcanic activity which periodically upsets life in the region.

Water is of vital importance, and for farmers the lack of it is a great problem. In the interior the farms are small, and produce wheat, olives and vegetables. One or two cows may be kept for milk, and sheep search for grass on the rough hillsides. On the more fertile coastal plains vegetables, flowers and citrus fruits are grown in valleys irrigated by water from rivers. On the slopes, grapes are grown for wine and olives for oil.

Sicily, like its neighbours Sardinia and Corsica, has a great many tourists attracted by the sun, sea and interesting history.

Cyprus and Malta

Cyprus, a large mountainous island of the eastern Mediterranean, resembles Sicily in its climate and terrain. It has a long and complicated history. After being part of various empires, it came under Turkish rule in the sixteenth century. In the 1800s it came under British occupation and later acquired the status of a colony. In 1960 Cyprus became independent but it is now split by disagreement between the Turkish and Greek populations.

On a visit to the island you would see picturesque, flat-roofed houses, interesting churches and mosques, and pine forests. Saint Paul and Saint Barnabas are said to have visited here and brought Christianity to the island. In the town of Paphos there are the remains of an ancient port and other archaeological sites. The birthplace of the legendary goddess Aphrodite is also near here.

A high proportion of Cyprus is under cultivation. Crops include cereals, citrus fruits, grapes, olives and tobacco. Mineral resources include copper and iron. Marble is quarried and used for some local building. There is a thriving textile industry and many of the people make lace and do embroidery. In winter it is possible to ski on the high, snow-capped mountains.

Malta occupies an important strategic position, linking the east and west basins of the Mediterranean. It is built mainly of infertile limestone, but has developed industries in farming and manufactured goods. Fishing too is important. Saint Paul also came here having been shipwrecked off its rugged coast. In early

times Malta had many different rulers; it became part of the British Empire in the 1800s. Later a British naval base was established here, but that has since closed. During the Second World War Malta suffered damage and was awarded the George Cross for her

Above The medieval walled town of Medina on Malta.

great loyalty to Britain. Today the Maltese islands are an independent republic.

Valletta, the capital, has some beautiful old palaces and churches. There is also a magnificent harbour in which large ships and small colourful boats can be found.

Above The Rock of Gibraltar stands like a fortress at the entrance to the Mediterranean.

Gibraltar and the Balearic islands

Gibraltar is a small, rocky peninsula and is joined to the Spanish mainland by a sandy and narrow stretch of land. The Rock reaches a height of 426 metres (1,398 ft) and the city is situated at the base of the western slopes. On the eastern side is a huge water catchment area which serves the whole island. There is no large-scale agriculture, and except for fish caught locally most food has to be imported.

Here you can see the huge harbour and the British naval base. There is also a much smaller harbour into which Lord Nelson's body was brought after the Battle of Trafalgar. Many battles have also been fought against Gibraltar itself, and in the Rock are ancient military installations including gun galleries. Ancient cannonballs and other war relics can also be found on Gibraltar.

The famous apes of Gibraltar live on the Rock, and it is said that if they should ever leave, Gibraltar will no longer stay under British rule.

Mallorca, Menorca and Ibiza, the three largest of Spain's Balearic islands, have recently put a good deal of effort into developing their tourist industry. Typical Mediterranean islands, the Balearics have more than three million visitors a year. Tourists arrive by sea and air, and new hotels have been built for them. Fishing and farming provide the food, and craft industries such as leatherwork or lace-making provide the souvenirs. In some places development has destroyed the unique landscape and beautiful beaches, so conservation is now needed if the character of these islands is to be preserved.

Above The gun gallery built into the Rock itself.

Above Young girls lace-making on Mallorca.

59

Algiers and Port Said

While all ports have the same basic purpose – to import and export goods – differences in history and trade lead to tremendous variations in the character of ports of the Mediterranean. Algiers and Port Said clearly reflect these variations.

Algiers is the capital city of Algeria and the largest administrative, industrial and business centre. Because of this it not only has the facilities of a port, such as docks and warehouses, but also offices, factories, shops and schools. In the Middle Ages it was the home of the Barbary pirates who terrorized sailors in the Mediterranean. After 1830 it was con-quered by the French, who built railways and roads. As the port grew, various industries were attracted to the city. Today, exports consist mainly of raw materials such as oil, coal and iron-ore from the interior, and cereals, wine, olive oil and wool from the fertile coastal plains. The goods are loaded on to ships in Algiers harbour. Algeria imports a lot of manufactured goods, but gradually, as industry develops there, more products such as cars and trucks are being produced.

In Algiers, life is as varied as the city itself. Around the harbour fishermen bring in their catches and dockers load the ships. Students

Right Busy street market in Algiers.

Below Barbary pirates board a merchant ship.

Above 'Bum boats' at Port Said advertise their goods to passengers about to sail through the Suez Canal.

from the university, speaking both French and Arabic, mingle with the tourists in the Casbah. This is a bustling, colourful place. Algiers really is a town where the old meets the new.

The harbour and town of Port Said were built at the same time as the Suez Canal in the late 1850s. The Canal cuts through the small neck of land separating the Mediterranean Sea from the Red Sea and Indian Ocean. Thus ships could avoid the long journey round the southern tip of Africa. The port is built on the western side of the Canal where it enters the Mediterranean. Initially only important as a coaling station, Port Said has developed a number of industries during this century, including cotton textile-manufacturing and petrol-refining. It was also linked to Cairo (the capital of Egypt) by rail and exported agricultural products of the Nile Valley. In 1967, during the 'Six Day' war between the Arabs and the Israelis, the Suez Canal was closed and Port Said evacuated. In the mid-1970s, when the Canal was opened again, it was necessary to expand and rebuild Port Said to make it an attractive place to live in and to recapture lost trade. Many of the inhabitants returned to the town and the Canal was widened and deepened. Today, however, the largest supertankers laden with oil cannot use the Suez Canal, which is shallow, and so a pipeline has been built to carry oil between the Red Sea and the Mediterranean.

These developments badly affected the economy of Port Said and in 1975 the govern-

Above All kinds of tankers, liners and cargo ships congregate at Port Said before making the journey through the Suez Canal.

ment declared it a 'Free Zone'. This helps industry because it means that duty does not have to be paid on imported goods, which are therefore cheaper. Port Said is now very modern. Small boats still bring in fish from the Mediterranean and farmers grow fruit and vegetables, but Port Said is also an important industrial and commercial centre.

Above Northbound tanker in the Suez Canal. This picture was taken from the bridge of a southbound ship waiting in the 'Ballah Loop'. The Suez Canal is too narrow for two large ships to pass each other, so special cuttings or 'lay-bys' have been made.

Oil and gas

You have seen that the Mediterranean is almost completely surrounded by land, and that it is connected to the great oceans of the world only by the Straits of Gibraltar to the west and by the Suez Canal in the south-eastern corner. Nevertheless, the Mediterranean carries an enormous amount of traffic, and not only between the ports around the shores of the Sea itself. It also forms part of an extremely important trade route from the United States and northern Europe down to Port Said at the entrance to the Suez Canal and from there to the Middle and Far East.

Since the Suez Canal was re-opened in the mid-1970s, the amount of shipping has greatly increased, and all types of vessels – passenger liners, cargo carriers and oil tankers – make use of the Mediterranean Sea. The area is now one of the busiest in the shipping industry. Even the extra-large tankers which sail from the Gulf, taking oil to northern Europe via South Africa, can return home through the narrow Suez Canal. This is because once they have unloaded their heavy cargo, they do not sit so low in the water and the relatively shallow Canal is able to accommodate them.

Oil is one of the most important commodities which is carried around and through the Mediterranean. Oil tankers do not all come from the Middle East and travel to northern Europe via the Straits of Gibraltar. They also trade between the many oil ports of North Africa and the European refineries on the northern Mediterranean shores.

The discovery of natural gas in various parts of the world has been vital in supplementing our energy resources. One of the largest loading ports in the world is at Arzew, in Algeria, North Africa. Highly sophisticated, gas-carrying ships travel regularly to northern Europe and the United States.

Below Tanker from the Middle East waiting to proceed through the Suez Canal with oil for Europe.

Other trade in the Mediterranean

General freight carriers also have an important part to play on the Mediterranean trading routes. They carry goods of all kinds from the southern European ports to many countries of the world. In recent years, much of this freight, or cargo, has been carried by 'container ships'. These move goods, already packed in giant containers which are loaded directly on to the ships from container depots at ports – a very quick and efficient method of getting cargo on board.

Cargoes which need to be kept at low temperatures, such as perishable fruit from countries like Israel, or fish travelling long

Below Pleasure craft in one of Sardinia's many harbours.

Above Ships under repair in dry dock on Malta.

distances, can also be carried in specially adapted container ships.

Many different countries border the Mediterranean shores, and it is important that the trading routes are kept open. So it is not unusual to find naval vessels belonging to the major powers throughout the world patrolling some parts of the Sea.

Fishing boats can also be found in the Mediterranean, of course, and make a much more attractive sight than the freight carriers. You can read about fishing on page 34.

With all this industry, it seems unlikely that there would be room for pleasure boats – for sailing dinghies, motor boats and larger cruising ships – but you can see these craft dotted about the Mediterranean all through the year, and particularly in the summer months.

Glossary

Algae   Seaweed and other similar plants.

Barbary   Old name for the part of North Africa comprising Morocco, Algeria, Tunisia and Tripoli.

Barnacle   A small shellfish which clings to rocks and the bottoms of ships.

Carnivore   An animal which eats other animals. A shark is a carnivore.

Casbah   The name given to the Arab quarter near the citadel of a North African city.

Container ships   Fast cargo boats which carry their goods packed in large containers rather than loose.

Continental drift   The movement of the landmasses of the world towards or away from each other.

Crusades   Expeditions of Christian soldiers to win back the Holy Land from the Mohammedans.

Crustacean   Animal (usually living in the sea) with a hard shell and many legs. Prawns, crabs and lobsters are all crustaceans.

Current   The flow of water in a given direction.

Desalination plant   Factory where salt is removed from sea-water so that it can be used for irrigation and domestic purposes.

Evaporation   The changing of water into a vapour, accompanied by cooling.

Herbivore   An animal which eats plants rather than the flesh of other animals.

Irrigation   Watering agricultural land by diverting rivers or by building dams, canals and reservoirs.

Kelp   The biggest seaweeds. Found below the low-water mark and growing up to a depth of 30 metres (100 ft) below sea-level.

Lampara   A long net used by Mediterranean fishermen to catch fish.

Limpet   A small shellfish which clings to rocks.

Mediterranean Ridge   An underwater mountain chain in the Mediterranean Sea, formed by Continents colliding.

Mistral   Violent cold wind which affects the Mediterranean areas of France.

Mosque   A place of worship used by Mohammedans.

Periwinkle   A type of mollusc, which is an animal without a backbone. Mussels, snails, whelks, octopus and squid are also molluscs.

Plankton   Tiny animals (zooplankton) and plants (phytoplankton) which drift in millions through the sea.

Pollution   Contamination of sea-water by dangerous chemicals from industry, oil spillage, and sewage or other rubbish.

Portulan   Decorative naval charts made for medieval seafarers.

Predator   Animal which preys on other animals for food.

Salinity   Saltiness. The salinity of sea-water varies a little according to the depth of the water and its distance from the Poles.

Sediments   Clay, sand and silt which collects on the sea floor and may become hard rocks. Remains of dead animals are also incorporated into the sediments.

Silt   Fine sand and fertile soil washed down to the sea by rivers.

Sirocco   Hot, dry wind blowing from North Africa over the Mediterranean and parts of southern Europe.

Thrust zones   Areas where the continents or sea floor have collided and pushed over or under one another – often the sources of earthquakes and volcanic activity.

Tidal range   The difference in a particular area between the level of water at low and high tides.

Tides   The rise and fall of sea level which usually occurs twice a day because of the attraction of the Sun and the Moon.

Twilight zone   The dimly lit region of the ocean from about 200 to 800 metres below sea-level. There is not enough light here for plants to grow.

Above The dramatic results of the major earthquake which struck Yugoslavia's Adriatic coast in 1979.

The people who wrote this book

Pat Hargreaves   Marine biologist, Institute of Oceanographic Sciences, Surrey.

Dr N. C. Flemming   Marine geologist, Institute of Oceanographic Sciences, Surrey.

Dr Robert B. Kidd   Marine geologist, Institute of Oceanographic Sciences, Surrey.

Margaret B. Deacon   Visiting fellow, Institute for Advanced Studies in the Humanities, University of Edinburgh.

Dr W John Gould   Marine Physicist, Institute of Oceanographic Sciences, Surrey.

Dr David Pugh   Marine Physicist, Institute of Oceanographic Sciences, Merseyside.

H. S. Noel   Journalist in fisheries and marine subjects.

Dr Andrew Campbell   Zoologist, Queen Mary College, University of London.

S. C. Pitcher   Teacher of geography, King's College School, London.

Alan Thorpe   Editor, *Shipping World*, London.

Index

Books to read

Angel, M. and H., *Ocean Life* (Octopus Books)
Cochrane, J., *The Amazing World of the Sea* (Angus & Robertson)
Keeling, C. H., *Under the Sea* (F. Watts)
Lambert, D., *The Oceans* (Ward Lock)
Merret, N., *The How and Why Wonder Book of the Deep Sea* (Transworld)
Parsons, J., *Oceans* (Macdonald Educational)
Saunders, G. D., *Spotters' Guide To Seashells* (Usborne)
Stonehouse, B., *The Living World of the Sea* (Hamlyn)

Picture acknowledgements